Dolphin Talk

How we can talk with dolphins in 5 easy steps

By Dolphingirl

Dolphin Talk

Copyright © 2014 by Dolphingirl

All rights reserved. Printed in the United States of America.

No part of this book may be used or reproduced in any manner whatsoever without written permission except in the case of brief quotations embodied in critical articles or reviews.

This book is a work of nonfiction, with research work and actual live expériences from the author, as well as wisdom from her nonhuman friends and Nature talking.

For information contact : dovejoans@gmail.com

http://www.dolphintalk.org

Book and Cover design by Dove Joans
ISBN: 978-1981959013

First Edition: October 2014
Paperback: December 2017

~ aloha ~

"You're only given one little spark of madness. You mustn't lose it."

~ Robin Williams

On that beautiful note, this book is dedicated to the misfits, and those who have felt any or all of the following:

> Animals are kinder than people
> Not from this world
> Separated and alone
> Wanting to fly, freedom
> Heart broken
> Raw and overly sensitive
> Misunderstood
> Disillusioned with life
> Overwhelmed and unmotivated
> Displaced and awkward with being yourself
> Incomprehensible desire to disappear into the seas
> Doubting Thomas... trust me, I was one of them
> Wanting to die... or essentially transform

This dedication along with Robin's quote is expressed because of the last year I experienced. I felt moments of deep loneliness, despair, and an utter sense of not belonging to anything, anyone, or any place. My saving grace was hearing the humpback whales singing in the sea off the coast of Maui. Their enchanting and calming songs helped me hold on, even though I didn't really want to...

I was weary of this earthly existence. Tired of trying my darnest to make something wonderful, be something wonderful, and too often, tossed on mee 'arse. They say everything happens for a reason. Well, I was full of reasons. My heart longed for more...

~Cetacean story~

Thankfully, towards the end of my vulnerable dark period, I was invited out onto a new friend of a friend's sailboat for the 1st time. As we cruised along, we excitingly watched the many magnificent humpback whales breaching off in the distance, as well as a few of them surfacing and passing by. It was at the end of "whale season", the beginning of March and my spirits were lifting.

When we were about 10 miles out from the coast, I jumped into the sea to have a listen and look around. I tried sending a *request* out to a humpback mum and her baby. *Asking if they would swim up to me,* and miraculously they heard me! Emerging through the deep blue seas, I saw the whales' forms taking shape closer and closer. So I waited in my fins and snorkel mask as they proceeded to swim directly up to me… within a mere 6-7 feet, as the new "baby" lead the way.

For what seemed like an eternity, the mum and I just gazed into each other's eyes. I felt her telling me, "Everything's going to be alright."

That peaceful and loving encounter was the turning point in helping me find my smile again, a *wave* back into my heart.

Table of Contents
(Food for Thought)

1. "aloha"
2. Table of Contents
3. Author's Note
4. My vision ~ paradigm shift and the cetacean brain
5. The awareness appetizer
6. The scrumptious salad
7. The marvelous main course
    The first step ~ ask
8. The second step ~ listen
9. The third step ~ be willing
10. The fourth step ~ giving up control
11. The fifth step ~ compassion in action
12. The determined desert ~ choices and avenues
13. News, networking, and non-profits
14. Special Thanks
15. Additional Books

~ AUTHOR'S NOTE ~

I come to *Dolphin Talk* with my curious and childlike nature, the same one we're all born with, as well as my investigative researching passions.
What makes a scientist or dolphin expert?
Studying, asking questions, and searching for the truth.
As an explorer of *interspecies communications*, one can not help but share their own personal stories, dolphin encounters, and living experiences.
I don't have all the answers. I'm merely sharing the wisdom from Nature and animals. Here you'll find suggestions to *waves* of life that's enabled me to "*talk* with the animals," especially my dear friends, the dolphins.
How did I get here?
So apropo when heard from a "Talking Heads" song. Mainly it's been through the cracks in my heart. I say that because of the brokenness I endured growing up. Unpredictable moments of physical, mental, and emotional abuse... not a constant daily threat, yet enough volatile surprises to cause nail biting, overweight issues, low self-esteem, shyness, speech challenges, and a looming dread instilled when a certain car was parked out in front of our home. Those transformational experiences and a "revelation" in my early teens propelled me on a quest for the meaning of life, like searching for 'The Holy Grail.'
My *dolphin epiphany* occurred in 1977 while listening to Roger Payne's *"Songs of the Humpback Whale"* and opened me up to a new world. After devouring all the books I could find on "cetacean communications" at my Mum's city college library, I announced to my

parents I was going to do "*interspecies communications with dolphins.*"

And that was that... well, kinda. Let's just say I *dove* off into the deep end.

What I instinctively knew as a child, I still believe in my "adult-like" life,

"Why not learn to communicate from the dolphin's world, their perspective instead of humans imposing our languages on them?"

There might be thoughts I share that may go against the "status quo" of perceptions, yet many years ago, humanity believed various notions like "the world was flat."

So, if something I say opposes your "beliefs", then maybe jump in...

Explore, ask questions, and be curious about your own truth. The lens in which we "sea" the world might become the animals that are *seeing* us.

All I know is that dolphins have been my family when I was broken, lost and didn't have the humanly support I needed at various times in my life. That heart support we often receive from a beloved pet, a family member that understands us without having to explain or prove ourselves.

I can also say from my experiences that animals have feelings, thoughts, memories, and attitude (plus they see in color.) They are often the conduit of their human family's feelings, yet more on that and other animal encounters in an upcoming sequel, *My Porpoise in Life*.

This is my 1st attempt in trying to describe what I thought most people could do naturally, *talking with animals...* then I found out it wasn't so common.
I tend to be a private person, yet thoroughly enjoy community involvement and sharing among close friends. For the most part, I've kept a majority of these animal stories to myself over the past 30+

years. When I did share a few of my experiences, I was either met with disbelief, ridicule, or both.
Out of curiosity, I once attended an "animal communications" workshop back in the mid 1990's, and ended up "*reading*" the teacher's companion pets from photos, because I got the "whole package."
When *tuning* into animals, a human can have 3 "interactions" ~ Hearing them, Seeing pictures, or the "whole package", Hearing, Seeing, and Feeling them. The whole package is like being a part of a movie, except it's real.
A present time happening of being in that animal's body, walking in their paws or swimming in their flippers. For whatever reasons, I've often been the reluctant animal *talker*. Each time it still feels like a gift-wrapped present, a divine surprise!

I've also been the type of person who can feel another person's (and animal's) pain. On many levels, I wished I wasn't so open... because I've been used as a filter, conduit, and scapegoat when trying to assimilate into modern society. I'm now "sea-ing" the interactions as *I'm a bridge*, possibly to other worlds to assist, support, and connect. Native cultures like the Native Americans Indians, Hawaiian Islanders, and the Australian Aboriginals have influenced my perspectives into "seeing" and "hearing" life in new ways, including the invisible world. If there was a word for it, it would be "dimensionally."
Honesty, there's been moments I've doubted my natural abilities, our natural abilities to be in "communion with animals", and really all of nature *talking* with us.
To remedy that, there's a saying, more like a mantra I now say to myself as I enter the ocean for a swim or snorkel, usually solo.... yet not for long, as nonhumans (animals) appear!

~ "I am One, you are One, We are One" ~

For those skeptics ~ I've heard many "New Age" notions, yet question many.
What I do believe in is more along the lines of *The Divine Age*, God speaking to us through Nature.
The Divine Age is something I've made up to describe God *talking* with us... through animals, through Nature, and sometimes very rarely through another human being.... mostly children, adults that are challenged, or individuals dealing with a terminal illness... ones I've had the honor to meet and share a bit of their lives with over the years.
"*The Divine Age*" is a timeless thing. It's crosses dimensions and has no boundaries. It's heart and mind expanding, in the naturalness of being, in just being!
An Australian Aboriginal would probably describe it as *Dreamtime*... a living and breathing connection that all of life is actively engaged with, whether we're aware of it or not. It exists like we exist, and every choice we

make is affecting the *dreaming*. In their sacred *dreamtime* paintings you'll see the dots and wave patterns of life's connections with nonhumans and humans alike.

~ A Bundjalung Aboriginal Elder, Lewis Walker painted and created this "gift" for me.
It was quite an emotional honor to receive it. With his hands outstretched, Lewis said, "You are doing custodian work for our brothers and sisters of the sea" ~
(NZ Maui's dolphin campaign)

Ever wonder why Indigenous people around the world have animals as their spirit guides?
I believe they are meant to guide humans with our spirits, not only in metaphysical terms, yet also in our thoughts, feelings, and actions.
I say that because if one observes animals in their natural state, their focus tends to be what they want, what they need, or just being.
Humans are the only animals that focus on what they DON'T want, what they DON'T need. Back in 1987 a wild dolphin told me telepathically *"Fear and anger are not in our vocabulary. We only recognize those frequencies from humans."*

NOTE: Want to know why it was the DOLPHIN'S thoughts, instead of mine?
3 reasons: #1 ~ Original thought. #2 ~ I wasn't thinking those thoughts prior. #3 ~ That clear idea was smarter than my current thoughts or beliefs.
Is "fear and anger" a gift we want to continue giving to our fellow nonhuman friends and to each other in this fragile yet magnificent world?
The Chumash Native Americans call the dolphin "a'lul'qoy", which means to go around, to protect and to go in peace.

~ Here's a charm of "celestial dolphins", a mini-replica by modern sculptor Bud Bottoms of an ancient Chumash painting found on a ceiling of a cave ~

~ **My vision** ~
*How do we shift humanity
into 'the dolphin wave' of being humane?*

My desire is for people to speak up and care for animals, acknowledging them as the remarkable nonhumans that they are.
So, I'm offering this *bridge* to individuals and communities around the world in hopes to unlock our *remembering* through the most sought after nonhuman in history, the dolphin. So "acts of kindness" become the main *wave* humanity rides with animals, Nature, and ultimately ourselves.
For instance: A big part of my vision is to help transform SeaWorld from a Circus to a "Cirque du Soleil", where natural sea-pens become the only habitats (verses cement tanks) for the now captive orca, dolphins, and whales, to then be re-introduced back into the wild and reunited with their sea-families.

Basically, bringing an end to capturing wild cetaceans for human amusement.

A "paradigm shift" can happen if we look upside down... accessing a window into the cetacean's brain perspective.

What am I saying?
What if our language flipped and that actual movement helped you and I in transforming, our 'worldview' forever altered?

~Cetacean views~
Have you ever seen photographs of sperm whales "supposedly sleeping" in the ocean? The pod will assemble in a formation where they're all upside down. Why is that?
What about trying to hear humpback whales singing their songs in the sea? Approaching it acoustically with our naked human ear, one of the best oceanic concerts can happen when one is upside down with our heads towards the sand. Virtually a headstand, or *"inversion" in the sea.*
Now let's take our language with the same "inversion" concept and use the word "Me" of the "me-mentality." When turned upside down the M in me, becomes a W, as in the word "We."
How can "We" apply this language shift in *our world actions*?
If we take the example of the yearly inhumane mass slaughtering of dolphins and whales in Taiji, Japan, what happens in the shifting of "me" to "we?"
The "Me-mentality" ~
A dolphin/whale is sold as food at $500 a piece, the meat contaminated and full of poisonous mercury for human consumption, harming humans.

In "dolphin trafficking", each pristine dolphin/whale is sold at $150,000 for humans' "amusement", and only a few people profit.

World shame with "dolphin families" separated and brutally murdered, plus claiming "tradition" while the majority of Japanese people don't engage in the sales, the profits, or the cruel actions. Side note: This "tradition" started in the late 1960's.

(Similar inhumane actions are currently happening in the Faroe Islands and the Solomon Islands).

The "We-mentality" ~

"We" recognizes the connection between cetaceans and humans.

"We" becomes aware of the nonhumans' families, bonds, and *waves* of living.

"We" chooses to stop the harmful humanly consumption, in meat and captive venues.

Industry for profits become "Eco-tourism," supporting wild and free rather than captivity. When **we** hurt nonhumans, **we** hurt ourselves. When **we** are saving nonhumans, **we** are saving ourselves.

Taking this one step further, on a biological level when a human goes upside down, our 4 major systems are benefiting immensely ~ heart, lungs, immune (lymph), and brain.

So doesn't it behooves us to think and act "upside down," advantageous all around?

Benefits of inversions:

http://www.lovelifesurf.com/benefits-of-yoga

"We" have the choice to make this shift, so why not make it together?
In 2011, I created the *Free Morgan PSA* (an educational video of activists coming together to help free a captive female orca, http://youtu.be/9zxZeO7IG4Y). A statement in the film from Jean-Michel Cousteau states, "We have learned a long time ago that marine mammals in captivity are like humans in jail."

Isn't it time we play our monopoly cards right by giving the dolphins and whales the "get out of jail" *free-dom* card?

Note: This can also be applied to all unnecessary "captive venues" used for humanity's entertainment ~ circuses, zoos, hotels, swim-with pools, and **not** including rescue programs where animals have been harmed, usually from humans and are needing our help to survive and thrive.

Imagine "sea-ing" into a dolphin's world... as if we put on a snorkel mask, and swam with an orca in the wild. Would we allow ourselves to be transformed?

*This statement was conceived from 2 experiences, one in a dreaming state and one in waking life. The dreaming one was with an orca pod.

Dreamt of orca, all of them coming in...

I went down to the ocean shore to meet them, yet I was in street clothes as I proceeded to get wet in the waves.
One big male orca called out to me, and his sound reverberated my body. They were a family forming a circle near the shore break. While I was walking closer to them, the tide was going out and I came upon random snorkel masks washing up onto the beach. As I picked one up, a realization "transfer" happened from the orca to me, they were confirming for us to "see their world", yet it seemed that humans had discarded or left these masks in the sea, like trash.

The waking occurrence happened while snorkeling off the island of Maui a few months ago. I was creating a new "sizzle reel" *invitation* for an eco-adventure documentary, *I am Dolphin*. While among the schools of fish, an ancient sea turtle appeared along the coral rocks happily feeding on the sea kelp. I introduced myself and this is how our conversation went.

Hello, I haven't tried talking to a turtle before, would you like to talk? PAUSED for a bit, *I'm wondering if you guys have names, like we humans give ourselves*? Sending it mentally/telepathically and then waiting by

turning my attention back to what was all around me ~ fish, ocean currents, and the present moment.
Out of the blue I heard "*Bruno*" pop into my head. I laughed into my snorkel piece, and said *Really? I thought you'd have a Hawaiian name. My name is Dove, or you can call me Dolphingirl. It's great to meet you.* Since I heard back from him (permission), I thought it was ok to ask him another question, *How old are you*? He replied quickly back with a "*102.*" I answered back, *Wow, I'm heaps younger than you.* Seeing that Bruno was busy eating, I sent a message of, *If you'd like to share anything more with me after you're finished eating, please come right up to me so I know.* I turned completely around and went back to enjoying the various interactions among the fish.
About 6-8 minutes later, Bruno appeared right at my side with his head looking at me. I was pleasantly surprised, and then I heard from him, "*Let's Go!*" So we started gliding together out towards the open ocean away from the coral rocks. Then all of a sudden, Bruno stopped, pausing to look around as if he was waiting for something to happen. So I asked him, *Are we waiting for something?*
He profound reply, "**This is my world.**"
He was showing me HIS WORLD! The beauty of that statement sunk in as we ventured further out. Then I told him, *I'm feeling nervous about how far we're going out*, so as he dove down to the sandy bottom, I said to him, *I'll wait here.* I watched Bruno from above as he made sand "angels" with his flippers and belly, while a school of fish joined around him in an amazing formation. I said, *Humm, I'm confused. I thought you were a male?* He said, "*I'm rubbing my belly.*" He finished the bottom massage, and started upwards. I sent out, *Thank-you, thank-you*, and watched as he moved out into the majestic seas, his home. I turned back towards shore, and unexpectedly burst into

tears. That experience moved me somewhere in my knowing, unknowingly. It was so impactful that when it came time to compose the script for *I am Dolphin*'s new "sizzle reel", I used Bruno's *perspective*.

- *I am Dolphin!* ~ http://youtu.be/qeF6Cpphs3g *

Warning: entering this world and the world of nonhuman animals might lead to life changing side effects... Effects such as: speechlessness, gratefulness, uncontrollable silliness, moments of unexplainable tears, kindness, heart expansion, wonder, waves of peace, and overwhelming feelings of "life-connections."

"If you want to find the secrets of the universe, think in terms of energy, frequency and vibration."
~ Nikola Tesla, Serbian inventor and engineer

Before we get into the delicious main meal of the "5 easy steps", let's talk about the world of *Cetaceans*, dolphins and whales.

~ The awareness appetizer ~

#1 ~ Cetaceans have been around on Earth for 40-50 million years and have learned to adapt beautifully to their oceanic environment. What have humans done with their environment? Modern homo-sapiens have been around 100,000 years, and over the centuries humanity has mostly been trying to make the environment adapt to us. Note: This notion does not apply to indigenous cultures that continue their practices of living in communion with nature.

#2 ~ There are 36 dolphins species, 32 marine and 4 river species. The biggest dolphin is an orca, the smallest, rarest, and critically endangered dolphin is the New Zealand Maui's, with an estimated 55 left on the planet.

#3 ~ A dolphin's brain relative to their size is bigger than humans. Placing our 2 brains side by side, we share many similarities... though a dolphin's brain is more developed in the ancient part and percentage wise, accessed to a higher degree.
Humans and dolphins also share a 4-chambered heart... a very powerful heart intelligence. Did you know that a heart emanates a magnetic field 5,000 times more powerful than a brain, and has been measured 6 feet out from a body? It's in constant communications with a brain, over 40,000 neutrons transmitted and communicated daily. Now that's a flutter of *heart talk*, and happening within both species.

#4 ~ Dolphins share highly complex social structures and family dynamics. Studies have shown that orca males in the wild will not leave their mother their entire lifetime (sounds a bit like a Jewish or Italian family... no offense intended, I'm half-Italian.)
Dolphins in the wild PLAY 85% of their day. Humans in comparison?
That percentage is dramatically lower, unless you're blessed to be around fun, curious, mimicking friends, family, and co-workers that have a very *CHEEKY* nature! Note: *CHEEKY* is an Australian/English saying that means, "mischievous nature, and often leads to *taking the mickey* out of someone."

#5 ~ Dolphins are sonic nonhumans. They see, play, live, communicate, and survive with sound. Their sound can create pictures, telepathic and holographic. The cool thing is we share a similar communication process, receiving/sending information through our jaws and transmitting the information to our brains to form "pictures."

The dolphins take it up a notch, so that when you're in the water with them, they can "sea" right through you, as in *x-ray*. In speaking terms, humans talk in a combination of analog and digital, and dolphins talk digital... like computer technology (sending *wireless messages.*)

This sonic technology and intelligence is a priceless gift to the human race. On the other hand, human's sonic testing for oil exploration and military testing are continuous "death songs" for our cetacean friends.

"Dolphins careth for man and enjoyeth his music."
~ Aristotle

If one could describe this book in "sound", it would a combination of musical songs... "Love Me" by Katy Perry, "Songs of the Humpback Whale" by Roger Payne, "Ocean" by John Butler, and "You and Me" by Pink and Dallas (You+Me).

In the style of a movie, *Dolphin Talk* would be a beautiful blending of *Avatar* by James Cameron, *Brother Sun Sister Moon* by Franco Zeffirelli, *Lucy* by Luc Besson, and *Whale Rider* by Niki Caro.

Ready to dive in and explore...

"Dolphins may well be carrying information, as well as functions critical to the regeneration of life upon our planet"
~ Buckminster Fuller, early environmental activist and futurist

~ The scrumptious salad ~

Acknowledge and appreciate ~
How many animals do you connect with in 1 day? Maybe try an exercise of *acknowledge and appreciate.*
For example, within 45 minutes of a Monday morning I *acknowledge and appreciate*d 1 lizard, 3 butterflies, 2 doves, 2 mocking birds, and a very curious dragonfly. The day before I remember and interacted with 3 dogs, 4 cats, a group of doves, 14 sea turtles (along the side of the bike path bordering the coast), a mongoose, schools of fish, and 3 shy yet

curious spotted eagle rays. Today I was greeted poolside by my usual swimming buddy, a dragonfly, and then a new larger one appeared *talking* and hovering over me. Also during the day I heard various birds outside my window, and then on my sunset snorkel, 4 spotted eagle rays appeared with a mum coming within 3 feet of me and becoming perfectly still so we could have this beautiful eye contact together.
It can be a simple "hello", "thank-you", or both. Today was an added WOW!

Taking a few moments in recognizing their presence and appreciating the reality that these nonhumans make our world a better place, a happier place. A world full of sounds, beauty, and wonder.
It's a present thing. Being "Presence" = PRESENT (a gift.)
Hasn't there been times in our lives when we've taken for granted the simplest of things while they're staring us right in the face, or touching our hearts? Begin doing this "focus" with animals, and this can easily be applied to our fellow human beings. *Gratitude is the attitude* to transforming challenging situations, and appreciating the moment.

"If all beasts were gone, men would die from a great loneliness of spirit, for whatever happens to the beasts also happens to the man ~ All things are connected." ~ Chief Seattle

Perception and Perspective ~
"Just because humans call them animals... doesn't mean we need to underestimate their abilities, or sensitivities" ~ Dgirl
"To underestimate somebody is not to recognize or understand their potential, their ability, talent as well as other aspects of the real person." ~ Dr. R. Luxemburg
Humans often use the terms of "He/she acted like an *animal*"... in ways that have implied aggression or out of control behavior.
Have we asked ourselves, what is humanity's true nature?
I imagine it would be along the lines of protecting, caring, and serving (through loving choices), and to

be a part of something greater than ourselves, to "co-create" (another form, "pro-create.")

If dolphins don't have fear or angry in their language, and only recognize those negative carrying vibrations from humans... are there *waves* for humanity to change?

Move closer to a dolphin's perspective?

What would happen if we took away our negativity and fear?

I've heard from Kahunas, sacred Hawaiian wisdom teachers that in the "old ways", ancient Hawaiians didn't have "negative words" in their vocabulary. If situations arose with aggravated energies, they practiced emitting sounds to dispel, rather then use their language in negative charged ways towards one another or in their environment.

Maybe this ancient wisdom can be applied to modern man's daily choices?

Conscious and unconscious ~
Dolphins (and whales) are conscious breathers, which means they are constantly in a state of "awareness." Every moment, cetaceans are choosing their breath, inherently a conscious choice.
Ric O'Barry shares a story in the movie, *The Cove* where "Kathy", the last "Flipper" dolphin *chose not to take her next breath*. That profound experience transformed Ric from a dolphin trainer to a dolphin activist, just from that 1 one precious breath. That's how powerful our breath is. How would humanity be, if we ALWAYS had to be conscious of our breath? Don't we often take "breathing" for granted (unless one has a respiratory challenge), especially when we slumber?
When you think about it, hasn't the ocean chosen us? Humans breathe naturally, one could say unconsciously. In essence, our breath has chosen us. Meaning, it's courtesy of the oceans we are breathing.

Ocean Wisdom ~
An important well-known fact is that the ocean gives vital life to our planet, the air we breathe and the water we drink. Ironically, the *health of the oceans* is not on everyone's priority list... let's explore that for a moment!
Like a mother's womb is to a baby embryo, the ocean is to our planet Earth... all living species, all life needs the sea to survive and thrive.
Like a mother's womb carrying and transferring the essential DNA to a child, the ocean carries and transfers the essential DNA to all life forms. Every breath we take is given to us from the oceans. Every sip of water we drink is because of the oceans. By *breathing* we are sharing the planet's life force, our ocean's DNA with every living form.

Humans are not the select species doing this, all living things are doing this.

So the question could be...

"Are we ever far from nature when living in the hustle bustle of cement city living? Nature says "No."

And the reason why?

It's because we're carrying Nature with us in every breath we take.

The ocean has been telling many of us these things that "science is trying to prove and has in sacred geometry's mathematics."
It's quite basic if you look at the ocean ratio to the planet and human beings water make-up. We mirror our blue planet's ratio of water to body mass... more than that, we are carrying the DNA of the ocean in our cells from million of years.
That's how we are connected to the sea, and our aquatic friends.
Many humans have momentarily forgotten this "ancient sea knowledge"...
I call it *the dolphin memory movement.*" It's in our DNA, our cells, and in the breath we are breathing, for each of our breaths is an ocean wave.
All of our cellular memory is stored in our DNA. Our memory is made up of information, and science says it's a form of energy.
Like matter, energy can't be destroyed, so our shared cellular memories are infinite.
This "life line", this DNA coding is in each of us, because all life forms share a *golden ratio* that's makes up our DNA. One can find it in a conch shell, a sunflower, a pinecone, a dragonfly, the tail of a seahorse, a windstorm, the curl of an ocean wave, Saturn, the Milky Way, and in us.

Ultimately, our shared DNA is carrying a complete blueprint of our universe from one generation to the next.

As I mentioned before, in our BREATH (air) we are sharing the DNA of the Ocean, thus the vital connection to life on Earth, and the Universe. It's no wonder indigenous cultures hold sacred, "water is life."
How do we know?
Our connections, cell memories, and values are recorded through a design pattern in life, referred to as "The Flower of Life."

This is the same pattern that cetaceans share with humans, which we'll discuss in depth during the main meal.

How does this pattern of life have VALUE?
Well, let's take a look at the question, When do humans typically place a VALUE on something important?
Isn't it frequently when there's a high monetary value attached?
How about our oceans?
Throughout any coastal community, we often see the ocean becoming valuable and worthy in real estate transactions...
Would you like an ocean view? A question posed in a house for sale or even a hotel booking that accompanies a high price point.

So why aren't humans placing a true VALUE on the oceans by protecting and caring for them?

What is needed for humanity to practice this VALUE with Nature?

What if we placed our heart's intelligence and consciousness into the factor of communication and conservation?

Scientific studies in cardio-energetics are showing us that there's a daily and dynamic heart "neuron-dialogue" with our brain. Roughly 40,000 neurons a day of "talking."

What if we applied this tremendous force into valuing our planet's personal womb, the seas? If science says there's "brain cells" in our hearts, why not use them?

"Above all we must realize that each of us makes a difference with our life. Each of us impacts the world around us every single day."
~ Dr. Jane Goodall

~ The marvelous main course ~

Step 1 ~ ASK ~

Permission to enter their world, "Would you like to *talk*?"
Essentially asking for a friendship.
Another way to see it... sending out a *request* or invitation.

~Dolphin story~

~ Whale and Dolphin Conference, Kona, Hawaii, 1992 ~

In January of '92, I flew back to the States from Australia to attend the "Whale and Dolphin Conference" being held in Kona, Hawaii.
After hanging out with Ric O'Barry (former *Flipper* trainer and dolphin activist), Jacques Mayol (freediving world champion, *The Big Blue* movie is based on), Dr. John Lilly (pioneer researcher in dolphin intelligence, inventor of the "isolation tank"), Estelle Myers (founder of the Rainbow Dolphin Center), and other cetacean advocates, I was invited by a flute-playing friend to come to the city of angels, Los Angeles.
He claimed he cared for the wellbeing of the whales by using their incredibly haunting and complex songs with his music. During that period, I became a very disillusioned *mermaid out of water* with the overwhelming elements of city-life. I had been used to "Nature living" in Hawaii, serenity in the mountains of

Colorado, and the raw and simply beauty of Australia back in the late 1980's.

So to remedy my conundrum, I journeyed down to the Santa Monica beaches. They were relatively quiet and deserted, people wise, which suited me fine... giving me the chance to commune with the sea so I could ask my dolphin friends for advice. I sat myself down in the sand facing the winter waves, and sent out, *AM I meant to be here? Please show me a sign?* My simple "request" to them... mixed with a plea.

You see, I was homesick and missed my wild dolphin companions in Australia. (with past *talking* experiences, it was a usual 15 minute-ish time delay. That's how I typically *sensed* our mutual encounters, like this reoccurring scenario: *the moment I set my towel down among the beach dunes, a dolphin family would appear close to shore and we'd swim together side by side, with the youngest coming closest to me.*)

Now, back under the cloudy Southern Californian skies, my exhaustion took over and I fell asleep back on my jacket.

I awoke from a deep slumber to see the dolphins waiting directly before me. THEY CAME! I leapt up, and without thinking, stripped down to my favorite word, *au'natural* and entered into the waves. Was a tad bit nippy, yet I was too happy to be thinking about that reality when I was off to be reunited with my oceanic family! I swam out to meet them beyond the sea break and we played together until I noticed a harbor seal looking at me. Feeling somewhat self-conscious, the reality of where, how, and why sunk in. I glanced back to the beach just as the lifeguard's jeep was traveling by my discarded pile of clothes.

Oh my, I'm naked in the sea in a BIG metropolitan city!

I waited until the jeep passed again, and then swam back in. Shivering, yet elated and peaceful, I hiked up the hill to visit a girlfriend's place in Pacific Palisades and was greeted with "You're crazy, now get into a warm BATH!" No argument from me.

This drawing by actor Sam Bottoms was created for me after telling him of my *first dolphin encounter* in 1984 (similar to the 1992 one.)

Another example of "*Asking*" is common among humans during childhood.
Remember your first crush, the overwhelming excitement you felt?
Oh, and the longing for that day when that special person would acknowledge you, giving you that secret look back.
What did we typically do, consciously and unconsciously?
We intently focused on what we wanted to happen... with our minds creating pictures, and most often,

those emotions and feelings were generated from our heart's pitter-patter. What next? We *sent out* our desires hoping they'd "read our minds." These natural childlike qualities are eternal, and can be directly applied to *talking* with animals.

Trust that they will hear you, feel you... sense you. Then trust yourself to do the same with them.

Bottom line, be open to *believing* that animals have feelings, thoughts, and intelligence... from that place, the possibilities are endless.

"In every living thing there is the desire for love."
~ D.H. Lawrence

Step 2 ~ LISTEN ~

Be open.
The way we are with music we love, with intent and freedom. You know how you "lose yourself in it?"... almost void of judgments, taking in every beat, every sound, every pulse, *the everything*.

Listening, such a powerful word, yet so gentle...
What you're listening for is Nature's reply, by allowing.

Sound a bit like a puzzle? Not really, just takes a bit of practice of not having an agenda. Meditating, praying, observing, and daydreaming help in this area of getting into the flow, getting yourself out of the way, or just *tuning* in.

~Dolphin story~

Last year, I was invited out as a *newcomer* on an early morning "mermaid" group swim at a south end bay of Maui. The ocean was somewhat hazy as I entered from the black coral rock coast. The rest of the humans had already departed towards the pod of dolphins frolicking in the middle of the bay. Even though my nonhuman friends were present, I still felt hesitate about going in when a number of shark attacks had taken place over the past few months, due to early morning murky conditions when "mistaken identity" had occurred. My "doubt" said I was swimming "alone," yet I quickly dismissed those fears and turned my thoughts on the dolphins, *asking* for a lovely encounter. During my 20 minutes swim out, my caution was thankfully alleviated when I eyed a trio of spotted eagle rays gliding nearby.

Entering the oceanic arena where 50+ dolphins and 20 humans were hanging about, I was immediately welcomed by a mum and her companion. How did I know she was a mum? She was in the beginning stages of GIVING BIRTH! Those magical moments of her before me, with the tiny baby tail emerging out of her was beyond words.

In the excitement of the "gift" she was showing me, I left my camera tied to my bathers. Instead, that birthing *imprint* was captured onto my heart. We humans remained with our nonhuman friends in the bay laughing and playing for well over an hour, which can be rare with wild spinner dolphins. As we humans were tiring and about to swim back to shore, the birthing mum and companion returned before me.

This time the wee BABY was halfway peeking out… she had invited me into her process, and feelings of being a new AUNTIE covered me!

Alas, I had given my GoProHero3 to a friend taking pictures, so again that miracle of life became a "soul-print."

Back on the shore our humanly group talked about how long and wonderful it was that the wild dolphins were playing with us.

In my exuberance I exclaimed, "A MUM WAS GIVING BIRTH! Did you guys see her?" The response was a brushed off no. I found out later that the "privilege" of going out with the mermaid group was no longer extended to me.

Yet my feelings were and still are, nonhumans are naturally the ones extending the privilege to us.

"We need a new paradigm for our relationship with dolphins" ~ Ric O'Barry

Step 3 ~ BE WILLING to ~

Take chances ~ leaps of faith, you know what they are in your life.
The ones you resist out of "rationalization," justifying that the "risks" are way too risky, even though you light up with enthusiasm at the mere   thought of taking those steps. Change, like the seasons are *natural* in Nature.
Receive ~
Put yourself in a receptive state... by being thankful. By forgiving.
Also exercising, laughing, playing, doing something creative, meditating, singing, playing music, cooking, making love, hanging out with animals, or being in Nature.
If you're feeling overwhelmed, and are **feeling like** you are not near Nature. Hold on... take a few minutes or longer to connect with some form of WATER. Run some water over your feet, hands, or back of the neck (biggest areas of nerve receptors.) Take a bath, shower, or jump into a pool. Find somewhere quiet, which may mean a bathroom stall or even a closet.
Try taking a wet cloth with you. Hold it in your hands, place it over your face and around your neck.
In those moments just breathe... focus on your breath and picture things that make you *happy*. Or visualize WATER running over you, through you. Imagine the ebb and flow of ocean waves moving in you.
Reminding yourself that the ocean, even though it may not be close in proximity is the vital element giving this precious planet, all living life the necessary air to breathe.
Respectively, *Our breath* IS THE OCEAN! So Nature is never leaving you.

*Ocean/Water ** Our Breath ** Nature/Animals
(Humans and Nonhumans)*

NOTE: I was "edited" by a large dragonfly on this one. My sensing tells me the truth, yet an old belief came up, and I typed, "when you're not near Nature" The dragonfly flew directly infront of my face while I was doing the *undulating flippers* in the pool. Mind you, I wasn't thinking at all about this part of the book. I was actually admiring the clouds in the skies and the birds flying by when the dragonfly appeared. Loud and clear in my head I heard her say, *"Nature is always with you... in your breath",* and then in an instant, she gave me the picture of how I had previously worded the "overwhelm sentence" above. Like she had already read my book. I asked her, "How did you know my book when it's not even published yet. She replied with an easy explanation,

"It's already in the collective consciousness." Yes, WE are SHARING the same universal *Consciousness*!

With Hawaiians (Ha) and Maori (Hongi) traditions, sharing *breath* is an extraordinary honor. It is considered to be a "sacred exchange" between 2 souls with a direct line to God. "Ancient Hawaiians recognized that their breath was the key to good health and believed it possessed *mana* (spiritual power)."

Trust your instincts and flow with your heart ~
Four years ago in 2010, I was invited to be the photographer for a Sacred Hawaiian Kahuna workshop on the Big Island. I was also luckily allowed to participate in some of the groups' exercises. One of those exercises was in 10 minutes or less, "Draw what you'd love your life to be."
Without analyzing or rationalizing, I just let myself create.
    This is my drawing:

Maybe try this drawing/coloring exercise with friends and family.
It's a wonderful visual aid to SEA what your heart is really telling you.

"Not a shred of evidence exists in favour of the idea that life is serious." ~ Brendon Gill, writer for 60 years, The New Yorker

Step 4 ~ GIVING UP CONTROL ~

"A Biggy"
Understanding the "Flower of Life" *pattern* can help get you there, when you realize it is the KEY... Key to what?
The *key equation* to living and all life forms. Divinity moving through us, around us, and in us.
In sacred geometry, the "Flower of Life" is made up of 13 "connecting spheres", and the core of spirit, *consciousness* is creating these spheres, these *wave*forms.
When this merging or connection happens between each sphere, a form is born called the Vesica Pisces. This *sacred form* (fish... a symbol used for Jesus (love), and dolphin) has been highly respected throughout history and ancient civilizations.
"Many mathematical functions and geometrical laws come into play within the shape of the Vesica Pisces that that are not present with a single sphere."
The Vesica Pisces relates to the basic function and shape of waveforms and vibrations. This "form" *communicates*, as well as BEING the *electromagnetic wave.*
I spoke of music earlier, music is the harmonics of a *wave*, which actually CREATES this *sacred form*.
　　Recognize it in the shape of a surfboard?

This *electromagnetic wave created from life and spirit comes into reality when we talk about our own "magnetic field" emanating around our bodies.*
At the same time, we have the *quantum sea of energy*, or "Ether" resonating from our body... etheric energy is essentially an energy field, a broad spectrum of "*wave* frequencies." We humans are resonating to this energy, so do nonhumans, planets, and life as we know it.
Another way of saying it is each "body of matter" (each of us) are creating a musical pattern, a moving *wave* frequency. When we are healthy, our frequencies would be *harmonizing,* like a rockin' band,

or a symphony. This spectacular *spiral movement* is happening in our blood, in water, in air, and in the sap of plants. Natural magic!

~Exercise~

Would you like to immediately *FEEL* your very own electromagnetic field?

Easy, there's a mini-me version between the palms of your hands.

A dynamic and magnetic circuit field happening all the time, especially when you place a living item in between your palms and focus with intent, not intense.

Heart intent... quite different.

A prayer, greeting, or sign of gratitude in many cultures.

Have you heard of the "laying on of hands" stories creating miracles?

Note:

I did the healing arts for 12 years, first on humans, then on nonhumans.

In extraordinary ways, nonhumans are very special to work on because they know exactly what's going on with themselves, and can tell you. Very transparent. Humans on the other hand are amazing, yet tend to have many layers of stuff to sort through, often called "static" energy. Instinctively my hands would feel a person's "feelings" locked away in their bodies, and sometimes even receiving pictures of what had happened in their lives. The same way "animals talk." Energy *waves* we each naturally store in our "cellular memories", whether these "memories" are currently serving us or not. Meaning: Our cell memories might be hindering our natural flow of expression, thus possible needing a "re-boot" to our "cell programming." Sounds like your computer, right?

Many practices in holistic healing as well as neurolinguistics explore this.

"Neither a lofty degree of intelligence nor imagination nor both together go to the making of genius. Love, love, love, that is the soul of genius." ~ Wolfgang A. Mozart

So where do you imagine dolphins communicate through to us?

In an attempt to try and describe it ~ it's through our "Ether" and our heart's "magnetic field" measured out from our bodies, namely our *heart intelligence*.
What's the magic word?
Vulnerability.
What modern society often says to protect, hide, and not be (unless one is a famed "entertainer", and then that *way of being* is celebrated and rewarded.)

The synonyms of "vulnerable" from Merriam-Webster dictionary are: endangered, exposed, open, sensitive, subject (to), susceptible, and liable.
Well, that's the access point of communications. The *vortex* motion through our hearts.
Not our heads.
Yet our heads are automatic "invited to the party" because of the constant communications between our hearts and our heads, regardless if one is *aware* or *conscious* about the party happening.
Scientifically measured, the emotion of "love" is a *band wave* that vibrates at a higher frequency, whereas fear is long and slow.
Since *sound waves* affects matter, the higher frequencies like "love", actually creates the design patterns matching the "Flower of Life."

This symbolic "Flower of Life" is found in many faith based teachings and ancient knowledge, and becomes a natural bridge between science and religions... *communicating* to humanity the invisible forces at play.

Imagine ourselves as a song, each of us playing a "*wave* pattern" with our instruments. Our instruments are our hearts, our thoughts, our emotions, and our shared cell memories (DNA.)
Literally, by changing our "*tune,*" we can *talk* with the dolphins.
There's truth to the universal saying,
"*LOVE is the most powerful language in the universe.*"
One could say it even holds *water*...

"Integrity comes from one who has the courage to act from the wisdom of their heart." ~ Harold W. Becker, The Love Foundation

Wonder why the oceans and water affect humans so immensely?
The oceans are touching our magnetic field by the ocean *WAVES*, the same "*wave* form" that makes up the "Flower of Life."

And water is naturally moving in the same *spiral motion* as our blood, the "*vortex*" connection. Scientifically proven, our resonating heart's magnetic field, the one that's 5,000 times stronger than our brain intelligence is blasting and COMMUNICATING thousands of neurons back and forth to our brain, a water-bond connection is formed between the ocean *waves and our human hearts,* thus ultimately sending those harmonious *waves* back to our minds.
NOW add the element of our DNA. DNA acts as an antenna to light and sound on a cellular level. We naturally have *water* molecules surrounding our DNA, and it's our emotions, specially the feelings of LOVE (higher resonating frequency) that allows us more access to our DNA codes.
Scientifically, we have 64 codes, and in most humans, only 20 active codes (amino acids.)

Essentially, we are holding the on and off switch with a "tuning" element in our emotion of love.

So how is this flower-like pattern important in giving up control?
Once we comprehend that our "spirit" or consciousness, proven both by science and faith, is in constant communications to all of life 24/7, we can in essence relax into our electromagnetic field "*talking*", or "*songing.*"
Maybe even start trying to trust our *heart intelligence* in regards to living.

"I," the me then becomes "WE"

Now the notion of *power struggle* serves what purpose? *We are enough.* From that point on, there are many avenues to communicate with nonhumans.

"The love for all living creatures is the most noble attribute of man." ~ Charles Darwin

I love scientists for their groundbreaking research. I also love the healing arts and the power of faith, knowing that the "communications" are coming from our spirit, measured out in the creation of the "Flower of Life" (which our heart's electromagnetic *wave*-field and Ether *communicates circularly* in our daily connections.)

For ease, I've named it **the dolphin memory movemen**t in communications and have created a rough diagram to help illustrate these concepts, specifically in regards to *talking* with dolphins.

The energy flows of fields you see below are our hearts' electromagnetic fields *communicating*. Kinda like our personal access codes, our individual songs. Dolphins are relating in ultrasound holography. Short waves of higher frequencies or vibrations making 3-D pictures. I believe humans can do this with our hearts, this *picture-communications*. It's the invisible world of *waves* we're communicating with, and something beyond our audible "hearing range."

To venture into *the dolphin memory movement*, just imagine for a moment that the *sounds* from cetaceans, their *songs* and *communications* are the gateway into accessing more of our DNA coding, thus more consciousness.

So, with all of these 'natural life connections,' does the concept of "separation" even exist? Based in physics,

sacred geometry and not in philosophy, this becomes an untrue perception (thought *wave*) we may be carrying around.
I know I've struggled with this illusion when I've had lapses, those old "cell memories" we're choosing to listen to when we have forgotten our true nature.

"Everything is energy and that's all there is to it. Match the frequency of the reality you want and you cannot help but get that reality. It can be no other way. This is not philosophy. This is physics." ~ Albert Einstein

Step 5 ~ COMPASSION INTO ACTION ~

Simply put, *talking* with animals (nonhumans) comes from the same place, the openness and the innocence of a child. *Original innocence.*

How do we get back there?

By remembering. *We are "co-creating" with God on a daily basis by our very nature, the nature of life. Proof, if needed is in our shared DNA patterns.*

By playing. The nice games, please let go of the head games. Why cause harm when we innately have the power to *be kind*.

    Simplify. The word says it.

By forgiving. Example: You find yourself frustrated either over issues beyond your control or you're unhappy about being momentarily unfulfilled.
    What do you do? Without thinking clearly, you take it out on a close family member or a friend. Then, who's the first to forgive?
    Easy one... the dog! Doggie wisdom says *lick uncontrollably.*
    Dog's *Meaning* of *Uncontrollably* = emotions with love.
Hanging on to destructive thoughts, by choosing not to forgive can lead to hindering our heart's natural expressions, those radiant "*wave* patterns." This concept applies towards ourselves, as well as other relations.

By feeding yourself healthy *foods of thoughts*.
Also one could say, healthy foods period.
Also, when one thinks about consuming animal products, it's hard to ignore the fact that one is consuming the energy and memory of that animal. The pain, feelings, and emotions that nonhuman went through to become another person's food to consume.
    Confession:
    I've been a vegetarian most of my life, have made an effort to buy locale and organic
    products, including pleather or recycled leather (vintage shops), and non-animal tested "living products." Since recently moving back to the Hawaiian Islands and spending more time in "their world", I've given up fish. Something that I was

hanging onto, justifying that dolphins eat fish and I was a part of their family. Truth is, there are other choices, kinder choices I can make readily available everywhere. Now that saying, "Having a *clear conscious*" takes on new meaning for me.

If you love animals, then this choice naturally gives back to them the love, respect, and appreciation they deserve.

It's another BIGGY to consider on this planet we all share... acknowledging that connection.

Acts of kindness ~ Volunteering, doing a daily uplifting gesture towards a human, nonhuman, or community. Practice the giving and receiving exchange by being *in service*. In your chosen field of work, think about creating "win-win" partnerships, and *the boomerang effect* by a "pay it forward" mentality, in thoughts and actions.

The *waves* we send out, return back to us in greater force = *ripples!*

Allow yourself to "be"... YOU!

GIVING yourself a break! From self-imposed pressures and life pressures. Try giving yourself a "time-out" by acknowledging the qualities that make you unique, as well as the little accomplishments you do in a day.
    There's only 1 of you on this planet... you are divinely special!

Are people or situations bothering you? Bring your energy back into focusing on your dreams, your heart's desires.

Try allowing and accepting them to be them, and you to be you.

One could view it as a form of *honoring* each other for our specialties, like the flavors of roasted coffee or a good cup of tea (can also support each other's frequencies, our "*wave* patterns.")

Feel wiped out, rest... knowing it's a natural flow in life to have quiet or down time, like the ebb and flow of tides in the ocean.

Note: For extreme situations, where nonhumans and humans are being harmed, please speak up and act with as much kindness as possible.

"Be the change you want to see in the world. We need not wait to see what others do." ~ Gandhi in 1913

~ The Determined Desert ~
CHOICES TO MAKE + AVENUES TO TAKE

ADOPT PETS, INSTEAD OF SHOPPING FOR PETS

PLEASE GO WILD, INSTEAD OF SUPPORTING CAPTIVE DOLPHIN SHOWS AND PROGRAMS THAT CAUSE HARM OR EXPLOIT ANIMALS

SIGN PETITIONS AND WRITE LETTERS TO GOVERNMENTAL OFFICALS

CREATE A CAMPAIGN (your version of outreach)
To illustrate: I *co-created* a "Dolphin Day" with Jean-Michel Cousteau's Ocean Futures Society in Santa Barbara for the *Lets Face it Dolphins*, a photo-petition online campaign launched to help save the pending extinction of the NZ Maui's dolphins. Over 2012 and 2013, I photographed over 2,000 photo petitions for the global campaign.
Efforts are still currently needed to support this crucial dolphin dilemma.
In addition, directed, wrote, and produced a short awareness video, *Through the Dolphin's Eyes*:
http://youtu.be/eNCYo255Yik

Here's a collage of a few petitions I took in Byron Bay, Australia, Santa Barbara, California, Nelson, New Zealand, BLUE Ocean Film Festival in Monterey, California, and Nashville, Tennessee (with additional photos on the campaign site.)

SPEAK UP FOR "KINDNESS TO ANIMALS"

LIFESTYLE CHOICES AFFECT NATURE, ANIMALS, AND US ~ BECOME "AWARE"

USE RECYCLABLE PRODUCTS WHENEVER POSSIBLE ~ ELIMINATE SINGLE USE PLASTICS

SHOP AND SUPPORT LOCALE

"More than machinery, we need humanity, more than cleverness, we need kindness and gentleness, without these qualities life will be violent and all will be lost. Do not despair, the misery that is now upon us is the passing of greed. You, the people have the power to make life free & beautiful."
~ Charlie Chaplin in *The Great Dictator*

~ Networking, News, and Non-profits~

Heart Intelligence ~ *The Heart's Code* by Dr. Paul Pearsall

Shark Conservation ~ *What's Cool about Sharks* by Jim Knowlton

Animal & Nature Preservation ~ *Nature Aware* by Rick Wood

The Interconnection of all Life ~ *Gaia Calls* by Wade Doak

Cetacean Captivity ~ *Death at SeaWorld* by David Kirby

Links between naval sonar and deadly mass strandings ~ War of the Whales by Joshua Horwitz

Studies by Naomi Rose, Senior scientist at Humane Society ~ http://www.pbs.org/wgbh/pages/frontline/shows/whales/debate/anticap.html, http://theorcaproject.wordpress.com/

Hawaiian 'Kahuna' workshops~
http://www.aumakua.biz/

Descriptions of "The Flower of Life" in "Spirit Science" video: http://youtu.be/RHuvW7YaGjQ
Article: http://www.world-mysteries.com/sar_sage1.htm

Exchanging Breath: http://www.hawaii-aloha.com/blog/2012/04/16/ha-the-breath-of-life/

Benefits of the Ocean:
http://www.huffingtonpost.com/2014/09/12/mental-benefits-water_n_5791024.html?ncid=fcbklnkushpmg00000063

Our Connection to Water:
http://www.academia.edu/753837/Evolutionary_Water_Wombs_Seas_Tears_and_their_Utraquistic_Relation

"New Paradigm" with Dolphins:
http://articles.sun-sentinel.com/2014-09-16/news/fl-viewpoint-dolphins-hunting-season-20140916_1_training-dolphins-dolphin-hunting-dolphin-meat

"Songs of the Humpback Whale"
http://cetus.ucsd.edu/voicesinthesea_org/videos/videoHumpbackAcous.html

Captivity Petition to sign and share:
http://www.gopetition.com/petitions/save-angel-the-albino-dolphin-from-a-lifetime-of-hell.html

Write letters to Congress and the U.S. Department of Interior about seismic and sonic testing, as well as animal captivity. Urging for alternative, sustainable, and humane choices.
Mailing Address:
Department of the Interior
1849 C Street, N.W.
Washington DC 20240
Phone: (202) 208-3100
E-Mail: feedback@ios.doi.gov

"Never doubt that a small group of committed people can change the world. Indeed, it is the only thing that ever has."
~ Margaret Mead

Non-Profits:

American Wild Horse Preservation Campaign ~ http://wildhor.se/ZmOhiZ

Ric O'Barry's Dolphin Project ~ http://dolphinproject.net

Jean-Michel Cousteau's Ocean Futures Society ~ http://www.oceanfutures.org/

MelbournDolphin ~ http://www.melbourndolphin.org/

The Oceania Project ~ http://www.oceania.org.au/ www.soundcloud.com/iwhales

Great Whale Conservancy ~ http://www.greatwhaleconservancy.org/

The Kevin Richardson Wildlife Sanctuary ~ http://www.lionwhisperer.co.za/

Elephant Voices ~ http://www.elephantvoices.org/

Earthrace Conservation ~ https://www.facebook.com/Earthrace

Voiceless ~ http://www.voiceless.org.au/

Ocean Alliance ~ http://www.whale.org

PETA (People for the Ethical Treatment of Animals) ~ http://www.peta.org/

Surfers for Cetaceans ~ http://www.s4cglobal.org/

Sea Shepherd Conservation Society ~ http://www.seashepherd.org/

NABU (Nature and Biodiversity Conservation Union) ~ http://www.nabu.de/en/

The Nelson Mandela Foundation ~ http://www.nelsonmandela.org

OceansWatch ~ http://www.oceanswatch.org/

WDSC (Whale and Dolphin Conservation Society) http://us.whales.org/

"Let us always meet each other with a simile, for the smile is the beginning of love" ~ Mother Teresa

~ SPECIAL THANKS ~

To the Creator, all nonhumans and humans I've met along the way, and recently during the writing of this book... the dragonfly who helped me edit, Bruno with his sea turtle wisdom, Charlie the cat, spotted eagle rays, bees, geckos, and various wild birds. My dear and supportive friends ~ Ben and Jamie Conway, Dolores "Vegan" Ericksen, Roy "Tarser" Baddiley, Jim "Shark" Knowlton, "Joybug" and Va Barber-Hua, Dean "Spiral" Campbell, Jan Hendrix, Greg Dahlen, Dwayne Devries, Tony Barry, Paie "Sparkler" and Andre, Connie Fueyo, Ric O'Barry and the Dolphin Project, Andy Van Roon, Mark Olson, Mark Romero, "Bomer," Georgio Umholtz, Jean-Michel Cousteau and OFS, Holly

Lohuis, Ben "Nicolas" Starling, Shawn "Whisperer" White, Darmin and "The Legend of the Golden Dolphin", Tommy "Dog" Longaberger, Mark Franklin and the Oceania Project, Jacqueline and Sharon with MelbournDolphin, Xavier Rudd, Chief Leon Shenandoah, Stephen Snyder, Bill O'Malley, Bud Bottoms and the Dolphin family, Harry Rabin, Adrian Belic and the *Happy* film, John and Josephine De Luca, Ted Crockett and NFF, Nick Dantona, Paul Clay, Gary "maHALO" Campbell, Tricia Lopez, Billy Block, Dennis and the Sunset Kidd, Carly Alyssa Thorne, Sher & Be Baum, Christina "Seahorsey", Dean "Jojo" Bernal, Phil Roberts, Howie Cook, Becci and Daniel with Planet Corroboree, Mayor Simon Richardson, Lewis Walker, Steve "Omar", Pete Bethune, Rick Wood, Jim, LaDonna and the kids, Surfrider, Scott and Whitney Bull, Jacob Tell and Oniracom, Wade Doak, Kenny Young, Mary Jo Rice, Sara Wilcox, Dennis Moran, Minouchkeen, Emiliano Campobello, Manuel and family, Dr. Roger Payne, Randy "Daddy" Blevins, Buddy Winston, Bill Mastrosimone, Gershon and Michael with the Great Whale Conservancy, Peter Kurland, Mick Williams and GoBamboo, Ian and Lorraine, Darice and Charlie, James and Glenys Dorr, Dr. Horace Dobbs, Ady Gil, Byron's Medicine Wheel, Peter Schneider and Underwatercam, Taggart Siegel and the film *Seed*, Randy Scruggs, Raul Malo and the Mavericks, Carlos Garcia and Living Aloha Magazine, Alex Greene and Noah, Amazon's Kindle, IngramSpark, Barnes and Noble Booksellers, Apple Inc., Facebook, LinkedIn, Andy Davis and family, Tom Pollock, Charles Tuthill, Rene Naufahu, Todd Banks and family, Michelle Sheather and family, Caitlin Smith, Scott Palmer, Don Pickering, Doris Thomas, Ross Murray, Adam Shostak, Randy Lacey, Ben Hellwarth, David Michael Wieger, Bobby Braddock, Steve Taylor, Jamie Weissenborn and family, Andy Dunn, Heather and Ted, Dondi Bastone, Bob Christopher, Janiele and Fiverr, Mike and Mimi deGruy, John Balian, Ed Giron, "Bearhawk", Lloyd Smith, Ed Ellsworth, Dean Jeffreys, Constance Boardman, João Parrinha, James Morgan, Steph Marie and family, Sylvia Earle, Ramzy Sweis, Gina and Sal Tassone, James Kleinert, Barbara Maas and NABU, Wyland, "Coco-bean", Trigg Schaefer, Ocean123, Samy's Camera,

Steve Harrison, Vistaprints, Adam Hall and Earthkeeper Alliance, ROOTS of Canada, Dan Aykroyd, "Cell Food", Indigenous Tribes, and The Pollination Project.

And in deep appreciation of musicians, comedians, artists ~ Chagall, C.S. Lewis, John Muir, Arnold Newman, Pink, U2, Sia, "IZ", Katy Perry, Beatles, Empire of the Sun, Johnny Cash, Peter Furler, Third Day, Robert Redford, Nelson Mandela, Brigitte Bardot, Pee Wee Herman, Oprah, Ellen DeGeneres, Craig Ferguson, Sam Simon, Ryan Blair, Daniel Lamarre, James Cameron, Anderson Cooper, Paul Watson, Lucille Ball, Jim Carrey, Mike Myers, Peter Sellers, John Paul and Eloise DeJoria, David de Rothschild, Ben Stiller, Viggo Mortensen, Dustin Hoffman, Audrey Hepburn, Gerard Butler, Meryl Streep, Johnny Depp, Maya Angelou, Ed Sheeran, Allison Krauss and the Union Station, Kevin Max, James Blunt, Dwight Yoakam, The Coen Brothers, Lawrence Bender, Larry Ellison and Oracle, Sir Richard Branson & Virgin Oceanic, St. Francis of Assisi, and the lovely humans who's quotes I've included in this book.

An extra thank-you to my 97 years old "Grandmoo" Beach, my "Moo" with her unwavering love, my Dad, 2 brothers, sister, nieces, auntie, extended family, and dear "Lucky" for helping me be the person that I am.

"Remembering the "Beach" cell memory, 'love frequency'

~ A story ~
I few months back I was showing my "Grandmoo" Jean-Michel Cousteau's book, *The Charm of Dolphins,* and in her innocence she said, "Oh is that about becoming a dolphin? Then you should like that."
Sweet words of wisdom and truth.

~ Photographs copyright © Dolphingirl ~

~ Additional Books ~

I am Dolphin ~ www.IamDolphin.com
"Beneath the waves of holographic communications"
The science + stories + soul behind our aquatic connections Volumes 1 & 2
(publishing date, 2017/2018)

My Porpoise in Life ~ www.MyPorpoiseinLife.com
Interspecies communications, illuminating animal encounters and their "nonhuman perspectives"
(publishing date, 2018)

Shine and Be Kind ~ www.ShineandBeKind.com
A children's book, empowering the younger generation to be an important part of the "*kindness movement*", sharing *waves* to love our blue planet!
(publishing date, 2017)

Memoirs of a Mermaid ~
www.MemoirsofaMermaid.com
Transformational adventures of an unlikely suburban girl into the world of dolphins, sea-secrets, and the unspoken language of the heart.
(publishing date, 2018)

THANK-YOU AND MAHALO!

~ www.DolphinTalk.org ~

www.ingramcontent.com/pod-product-compliance
Lightning Source LLC
Chambersburg PA
CBHW040226220526
45473CB00001B/141